THE TECHNIC
OF MECHANICAL DRAFTING

A PRACTICAL GUIDE
TO NEAT, CORRECT AND LEGIBLE DRAWING

By CHARLES W. REINHARDT
CHIEF DRAFTSMAN ENGINEERING NEWS

SECOND EDITION

NEW YORK:
THE ENGINEERING NEWS PUBLISHING COMPANY
1904.

PREFACE.

In the following pages the writer has endeavored to give to the busy draftsman a thoroughly practical, common-sense guide to good mechanical drafting. No attention whatever has been paid to the mathematics involved.

Many draftsmen, after having carefully laid out their drawings, commit the error of slighting them in the execution; sometimes to such an extent as to render them all but illegible to the men in the shops. The constructing engineer frequently finds himself handicapped in a similar manner. After wrestling with the meaning of portion of a drawing, he finds that other sections or elevations do not correspond with the part first consulted. He has to make his own deductions, and in reading between the lines, as it were, perhaps finally arrives at some sort of conclusion regarding the correct shape of the parts outlined, or, as a last resort, goes in search of the draftsman. Such inconsistencies, if they may be termed so, unnecessarily increase mental labor in reading and interpreting a design, and it will be the special province of this book to try to eradicate them.

While deprecating any needlessly elaborate finish, the writer advises the use of just sufficient shading and finishing touches to render a drawing thoroughly comprehensible, and to preclude any possible ambiguity.

The illustrations in this work, quite a number of which have appeared in the columns of "Engineering News," are inserted to demonstrate the points involved, and have been reduced more or less in size so as to save space. They may, however, be copied by enlarging to about three or four times their linear dimensions.

It is of course to be understood, that this book is not a manual for beginners exactly; it serves its purpose best, where used by the draftsman who is familiar with the mathematical principles of mechanical drafting. It will then, as the writer hopes, prove a valuable aid to the rapid production of neat, correct and legible drawings.

New York, *December*, 1899.

CHAS. W. REINHARDT.

SYNOPSIS OF CONTENTS.

CHAPTER I. GENERAL INSTRUCTIONS.

THE drawing of any object upon either drawing paper, tracing linen or paper requires clean-cut, sharp outlines. It is a mistake to suppose that very thin outlines give an especially neat appearance to any drawing. Main outlines should always be drawn with the nibs of the ruling pen slightly separated, which "setting" will give a fixed width of line and allow an uninterrupted flow of ink. When refilling the pen, care should be taken to readjust it to the proper strength of line. To those unaccustomed to this somewhat delicate task, the use of a "lever" pen, which can be obtained from almost every instrument dealer, is recommended. This pen, as shown by Fig. 1, opens with a lever, and will, after refilling, yield the same strength of line as before. In regard to the ink to be used, the author would

Fig 1.

recommend nothing but the "waterproof" kind, a somewhat refractory medium to the uninitiated; but this ink will not smudge or rub off while being handled, a quality which the ordinary inks do not possess. After refilling, the pen should be tried a couple of times along the grain of a clean portion of the drawing board, or across the fingers of the left hand—a perhaps objectionable but always effective expedient—to insure a "starting" of the flow of ink. The actual work with the pen, owing to the rapid drying of the waterproof quality, must of course be commenced immediately afterwards.

A ruling pen with long, straight nibs will be the most satisfactory instrument for working with waterproof ink. The draftsman choosing a pen should open the same lightly and look through between the nibs against the light. The two outer curves, "a" and "b" (Fig. 2), should in a good pen be very flat, coming down to a long and very thin point, as illustration shows, and not appear as at "c" and "d." The ink should never be allowed to run out of the slot and gather on the outside of the two points, as will happen when the pen is screwed up so tight that the points touch, a pernicious practice indulged in by some draftsmen in the belief that the very finest lines can only thus be obtained. In reality the points of the pen cut into the surface of the paper and deposit the particles of ink rubbed off from the outside of the points, yielding thereby a faint, grayish, ragged line, which will not

Fig 2.

reproduce well either by blue-printing or photography. A very good, black and clean-cut thin line can be obtained by leaving the nibs just a trifle apart, which procedure will cause a free deposit of liquid ink that will invariably reproduce by either of the processes mentioned. The width of the opening cannot be given definitely; it should vary according to the style of drawing and the individual inclination of the draftsman. The instrument should be kept bright and clean and never be allowed to accumulate a coating of dry ink at the points. Wipe, or if necessary scrape off. All-around cleanliness forms one of the principal requisites for the successful draftsman.

A worn-down ruling pen can be put into proper trim again by the draftsman himself through the exercise of a little patience and care. The pen, which always wears off, as shown in Fig. 3, is applied to the oilstone, which every draftsman should possess, and the points ground off uniformly to the original rounded outline and of course dulled, as shown, by passing the pen over the stone, as if to draw lines upon it, first inclined, then gradually rising to the upright position. By judicious grinding of the outer sides of the two points the requisite thinness and knife-edge is finally restored. A pen may in this manner be kept in good repair a considerable length of time, until it is finally ground down so far that the thinning of the points would require too much time to be a paying operation any longer.

The paper to be used may either possess a fine-grained or coarse (egg-shell) surface; special recommendations in that respect would be beyond the scope of this work; tracing paper or tracing cloth can be obtained in excellent quality from almost any dealer. As regards the most suitable side of the latter material to work up-

Fig 3.

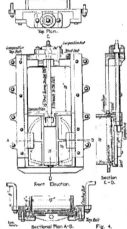

Top Plan.
C

Front Elevation.

Section
C-D.

Sectional Plan A-B.   Fig. 4.

on, the writer would advise the use of the smooth or inner side for various reasons, one of which is that on the back or unglazed side the lines will invariably become ragged or broken, as there is no continuous surface for the ruling pen to glide over. Erasures, furthermore, are almost impossible upon that side. The inner or smooth side presents a glass-like surface, which at first may not readily take the ink; a little brisk rubbing with powdered pumice, will somewhat deaden this gloss and admirably prepare the surface for the ink. Some tracing cloths abound in "leaky" spots, the ink soaking through at those places, especially where heavy lines are used. Such blots can be erased with a sharp penknife on the reverse side after the drawing is finished. Ordinary erasures on paper are usually made with the same medium, after which a good artist's rubber may be applied. Erasures of large sections of ink lines or blotches on tracing cloth can be splendidly effected by using pumice powder sprinkled over the parts to be erased; brisk rubbing with the tip of the finger, or perhaps a "circular eraser," while gradually replacing the discolored powder by fresh pumice will quickly clean the part in such a manner that it can be lettered or drawn upon without the least inconvenience; the cloth has retained its surface and will not blot after this treatment. Smaller areas, as, for instance, a short portion of a single line close to another, can be treated in a similar manner through a slit of proper size cut into a piece of tracing cloth placed on top, which covers and protects the adjacent parts.

In arranging the different parts of a drawing on a sheet, follow the ordinary common sense rule. Place elevation and plan in vertical projection ; if a top plan, place above; if a bottom plan, below the elevation. If an end elevation of the right hand end of the object is to be shown, project it from that end of the side elevation, etc. Enlarged details of some parts can, where properly captioned, be placed as "fillers" almost anywhere. The foregoing is to a certain extent illustrated by Fig. 4.

CHAPTER II.   OUTLINING.

THE visible lines which define the edges of any object represented should, after having been carefully penciled in, invariably be drawn full. For construction, "invisible" or projection lines, where such are desired to be shown, dashes about $\frac{1}{16}$ inch long with spaces of $\frac{1}{32}$ inch between should be used. Two or more parallel lines of that order placed close together, defining, for instance, the thickness of a plate, etc., should be ruled in so as to "break joints," as shown in Fig. 5. In a well executed drawing the dashes or spaces of those "invisible" lines should be uniform in size, a rule which is only too often disregarded, and as a consequence will cause a peculiarly "slipshod" appearance of any drawing, (see specimen, Fig. 6).

Center lines, axial lines, datum lines and lines of section should invariably be represented by dash and dot lines, the dashes twice as far apart as those of the construction lines, and the oblong dots

Construction Lines.

Center Line

16 to 20
Dimension Line

Alternative Line

Fig. 5.

A.

placed midway between dashes. Such lines should, for the sake of distinctness, always begin and end with a dot. Occasions may arise where the introduction of a line of this order, containing two or more dots between dashes, may become desirable for a specific purpose.

The writer notices that the average draftsman is especially careless as regards execution of this kind of lines, the dashes often being made so long and so irregular in size that a line of this order cannot always be recognized as such (see Fig. 7).

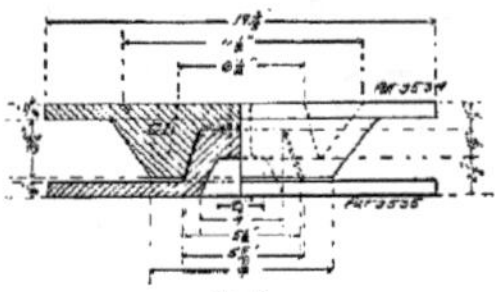

Dimension lines should consist of oblong dots or very short dashes, about 16 or 20 to the inch. A suitable blank space in the center should be left for the dimension proper. As a certain amount of practice with the ordinary ruling pen is required before long lines of that order can be drawn evenly, a "dotting pen" might with advantage be employed. Lines which indicate alternative positions of an object should consist of half length "construction" dashes. In regard to construction lines, and especially dimension lines, the ordinary practice differs from the rules laid down here, very thin black ink, pencil or red ink lines being employed to denote those lines. At the same time the tendency in using such symbols is to give construction lines a secondary character, and to make dimension lines least prominent. A drawing for photo-reproduction therefore, which is executed in pure black, as it properly should, certainly ought to contain such distinctive lines also

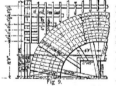

Fig. 6.

Fig. 7.

in that color. Where rather small objects are to be shown as "invisible," the rule given for the construction of such lines may be modified and the lines shortened as necessary, so as to outline corners and distinctive features, as shown in Fig. 5A. The first dash of projection lines proper, to which category also the abutting lines between dimension lines belong, should never be permitted to touch the "visible" or full outline of an object, as otherwise the continuity of same be interfered with, as shown in Fig. 8, with which Fig. 9, as corrected, is compared. Here we see that the values of the dimension and, of some center lines in Fig. 8 virtually have to be deducted by the eye in interpreting this drawing, whereas in Fig. 9 the same number of dimensions, etc., do not in any way interfere with the visible outlines of the object, which is allowed to stand out clear and distinct.

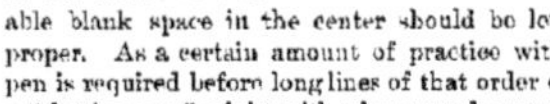

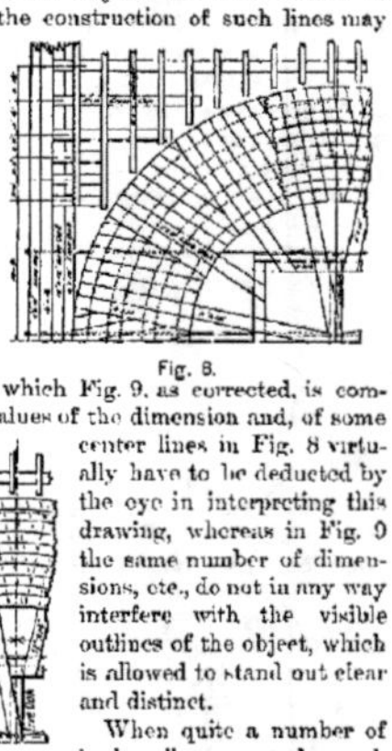

When quite a number of broken lines are to be used, making a rather confused tangle, it will also be advisable never to let a dash of such line cross solid outlines; this will have the effect

Fig. 8.

Fig. 9.

of keeping the latter more distinct, and will aid the reader of the drawing in easily forming a mental picture of the outlined objects in relation to the hidden or invisible parts. This point is fairly well illustrated in Fig. 12, which represents the rear end of a locomotive. The heavier outlines, which in that figure are shown to heighten the effect desired, will be explained subsequently. Notwithstanding the fact that the full outlines show the objects almost in relief, as it were, the "invisible" lines, while not interfering with the full ones, are not in the least slighted; the roof beams of the cab, the stays, fire brick arch and the rocking grate of the fire-box, the several levers with attachments, etc., show very clearly, as if through transparent material.

A good example of "invisible" and "alternative" lines contrasted is given under Fig. 10, a drawing of a folding car-step; the latter kind of lines indicates graphically the parts folded up.

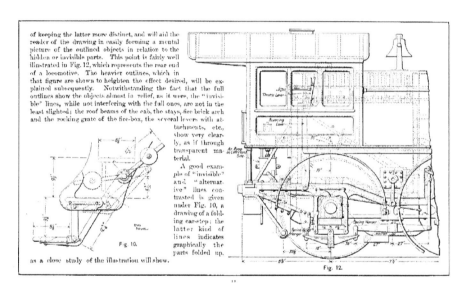

Fig. 10.

Fig. 12.

as a close study of the illustration will show.

In finishing a drawing first of all should be drawn in the curves that are to be connected with other lines. Then the horizontal lines or those nearly so may be drawn; and next the vertical and diagonal ones. Curved lines on any part of a drawing that are not intimately connected with adjacent parts may be inked in at the draftsman's convenience. Short portions of curves, which cannot easily be matched by the curve ruler, afford valuable practice in freehand drawing and should unhesitatingly be drawn in that manner, after making sure that the writing pen used for such an emergency is pliable enough and not too fine, so as to give the requisite strength of line. A smooth joining of curves with straight lines is essential (see Fig. 11). Small circles, such as designating rivets or bolt-heads, may in cases

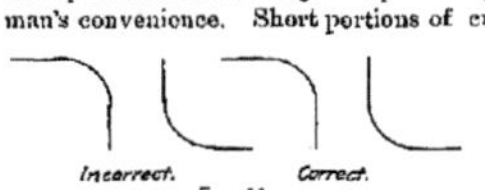

Incorrect.   Correct.
Fig 11.

be drawn last of all. Definite rules, however, to cover every instance cannot be given here.

After the outlines of the drawing have thus been secured, the projection lines are ruled in; the dimension lines are then placed between them, leaving suitable blank spaces for dimension numerals. If outline shading is desired, this work will be next in order, beginning again with the curved and horizontal shade lines, and continuing with the vertical and diagonal lines.

The lettering of the drawing should then be placed; care must be taken not to make the notations, descriptive matter and dimension numerals too small—a very frequent mistake. Good lettering and the manner of distributing same over the drawing materially adds to the neat appearance and clearness of any sheet; the reverse is unfortunately only too often noticeable. A competent draftsman should also be an expert letterer.

CHAPTER III. LETTERING.

FOR a thorough course in the principles of construction of freehand lettering for working drawings, the mode of procedure and sequence of strokes, etc., the reader is referred to the author's "Lettering for Draftsmen, Engineers and Students" (The D. Van Nostrand Publishing Co., New York). We intend in the present volume, however, to show a few suitable alphabets, (inclined and upright gothic) which adapt themselves to the lettering of almost any kind of drawing. These gothic styles, as shown, can all be constructed with a single application of the pen, are easily executed and look neat and business-like on a drawing. Where a draftsman has learned to construct those styles properly, he will find no difficulty after a while to do more elaborate lettering, as the fundamental shapes and proportions of the letters have become familiar to him, and are produced like ordinary carefully written or copied English script

The pens which are used for this style of lettering should yield a

abcdefghijklmnopqrstuvwxyz
ABCDEFGHIJKLMNOPQRSTUVWXYZ
3/4"; 1 5/8"; 3 7/16". *1234567890 & I II III IV V VI IX X etc.*
Inclined Lettering, Used for Descriptive Matter.

abcdefghijklmnopqrstuvwxyz
ABCDEFGHIJKLMNOPQRSTUVWXYZ
Condensed Style. & 1234567890 Extended
Upright Lettering, Suitable for Captions.

Fig. 13.

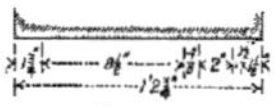

uniform strength of line. As regards the quality or grade of pens to be used the reader is again referred to the publication quoted above. The inclined lettering "cap and lower case" shown on Fig. 13 should invariably be used for all descriptive matter, such as dimensions, notations, etc. The upright style "cap and lower case" may be employed to emphasize names of principal divisions or to designate titles or captions of some portions of a drawing. The dimension numerals of a drawing should be placed boldly within the blank space left previously in the dimension line, or, space not permitting, set them outside and point by a dotted reference line to

Fig. 14.

the center of the space so designated (see Fig. 14). The two sides of arrowheads, which should either be pointing outward or toward the dimensions, should be made with a fairly fine, pliable pen in gradually increasing and finally decreasing strength of stroke, making sure of a symmetrical head and a good, clean point, as shown enlarged in Fig. 15. The last dots of a dimension line against the abutting lines might also be omitted, so as to aid the effect desired. The fraction numerals of dimensions should never be made less than ⅓ the height of the whole numbers; they should, in point of fact, be made extended, so as to seemingly make them appear larger, as shown in one of the following figures.

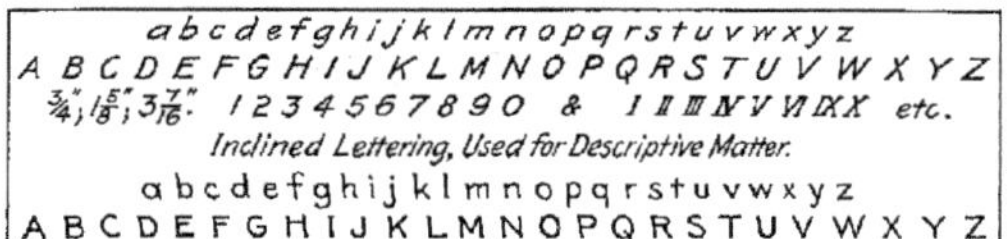

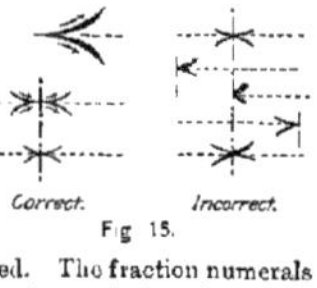

Correct.   Incorrect.
Fig. 15.

The wording of the descriptive matter (inclined style) should be well distributed; indeed, a good draftsman is able to combine the useful with the artistic aspect by properly arranging that kind of lettering. Just a word or two in

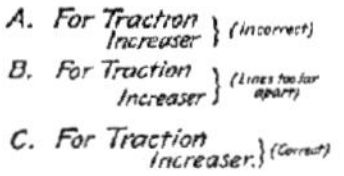

A. *For Traction Increaser* (incorrect)
B. *For Traction Increaser* (lines too far apart)
C. *For Traction Increaser* (correct)
Fig. 16.

regard to the arrangement of the lines of a notation: Either center them neatly, or, better still, stagger them; by no means, have them appear as in Fig. 16, A and B. Take care to get the different lines composing a note very close together, so as to facilitate reading.

In the author's "Lettering," mentioned above, the "extended" style of inclined and upright script is brought quite prominently to the student's notice. It may become at times desirable to spread a word or words over quite a large space : the single letters are spaced further apart, and the letters themselves drawn extended, so as to comfortably fill the allotted spaces. In Fig. 17, the correct way of extending spaces as well as letters, is shown, as also the one-sided method of only increasing the width of spaces without extending the width of the letters. A somewhat surprising effect is noticed when we compare the extended with the ordinary and condensed styles of lettering of exactly the same height, as shown on Fig. 18. Here the former type appears almost twice the height of the latter, and it demonstrates to what extent such styles may be utilised on one and the same drawing, for the purpose of giving to some parts more prominence than to others.

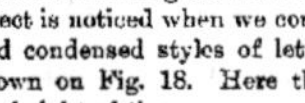

T r u c k *(Incorrect)*
T r u c k *(Correct)*
Fig. 17.

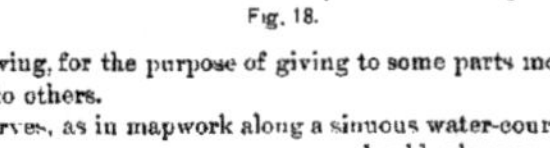

Line of Bluffs extending North and South.
NEW YORK STATE FOREST RESERVE.
Manhattan Suspension – Bridge.
Fig. 18.

Lettering on curves, as in mapwork along a sinuous water-course, should always be executed to conform with the different radii of the compound curves, that is to say, the vertical stems of such letters should be placed truly radial. Where inclined letter-

COLORADO RIVER
Fig 19.

ing is to be used to designate such curves, allowance will have to be made for the angle of such downstrokes with the vertical, as Fig. 19 illustrates. Where the curvature of a river-course is at points too sudden to admit of an easy continuous curving of its name along its side, such portions may be disregarded and the curves eased.

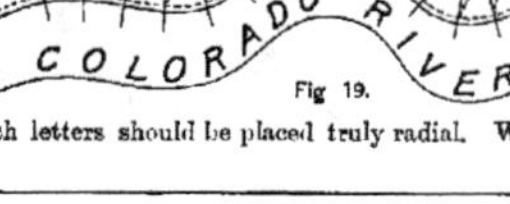

Fig. 20 contains a list of contractions and abbreviations which are most commonly employed in lettering working drawings ; as

6'0"Diam. Circumf. 12.7 ᶜᵐ Cu.ft. 1.4%Grade
3'0"Rad. El.±0.0 4.2sq.ft. 3.5yds. ∠=8°45'13"
12'7½"Q to Q (Out to Out) 10'0"C. to C. 4'3⅛B. to B. (Back to Back)
R.S., L.S., (Right or Left Side) H.W.L., M.W.L., L.W.L. (High, Mean, Low Water Level) 6"x8" Y. P. (Yellow Pine) W.Q. (White Oak) 6.5lbs. 79°F.
Rev.per min, R.p.m. (Revolutions per Minute) HP. (Horse-Power)
E.HP. (Electrical HP.) Eff.HP. (Effective HP.) B.HP. (Brake HP.) I.HP. (Indicated HP.) H.p., I.p., L.p. Cyl² (High Pressure, Intermediate P., Low P. Cylinders) M.e.p. (Mean effective Pressure) C.R. Shaftg (Cold Rolled Shafting) C.S. (Cast Steel) C.I., W.I. (Cast or Wrought Iron)
1½"ᵈ or 1¼"ᵈ Rod, 2L§ 5×3¼§ 1-6"L; 2-10"T§ 4-6"Z§,
1, §"Web Pl., Spl.Pl., Spl.L§, Diaph. (Diaphragm) Latt. Bar,
Lac'g L§; Kw. (Kilowatt) Amp. (Amperes) f(finish) t (turn)
Hydr. Gr. (Hydraulic Grade) Portl. Cem. Conc. 1· 2 : 4
(Portland Cement Concrete; 1 Part Cement to 2 Parts Sand & 4 Parts Stone)
Fig. 20.

some draftsmen are at times at a loss for correct or logical abbreviations, the writer thought it worth while to compile the list here given. Some of the abbreviations here given can, of course, only be employed where their use precludes any possible ambiguity.

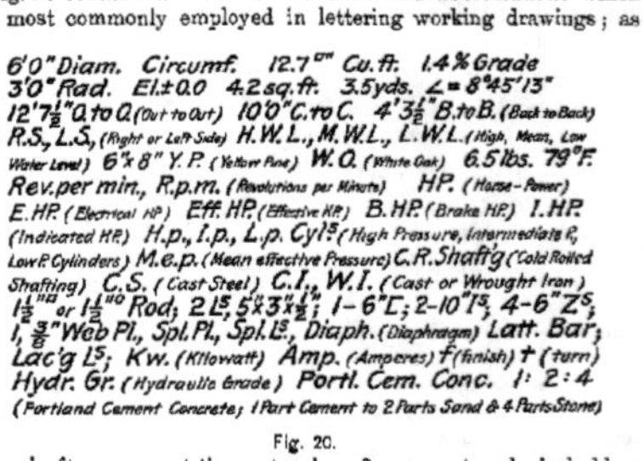

Captions should invariably be given ; every separate projection should be named distinctly in upright lettering, as, for instance, Plan, Elevation, Transverse Section, Enlarged Detail "C," etc. Not only the general appearance of the drawing is improved by using specific captions, but additional clearness is gained. The main title in the lower right hand corner of the sheet or at its base is placed after a border line has been ruled in. Make the border consist of single or perhaps double moderately heavy lines; the writer would not recommend the use of any ornate design for borders; they are as a rule antiquated and take much valuable time to construct.

In regard to the general title of the sheet, few, if any, directions will be given here, as almost every book on lettering in the market takes up this subject in a fairly thorough fashion. Let such a title be composed of simple, easily formed letters, preferably Gothic, made conspicuous enough for the main appellation of the subject, secondary in size and body for names of engineers, etc., and have it

supplemented by a mention of the scale of the drawing, lettered still smaller in size, with the actual scale attached. Underlining of words or notations on a drawing is bad practice and should only be resorted to after all other distinctions have been exhausted. The writer's book on freehand lettering gives reasonably fair examples of Gothic and Roman large letters suitable for titles, as also of actual specimens of titles taken from working drawings; while Jacoby's "Plain Lettering" may be recommended to those wishing to arrange their titles in a scientific and accurate manner with letters correctly spaced and proportioned. For ordinary purposes, however, the average draftsman is well able to space letters by eye only, and make quite a satisfactory general arrangement for a title.

CHAPTER IV. OUTLINE SHADING.

IN everyday practice this kind of shading is unfortunately not employed very often, and upon some kinds of working drawings its use would perhaps be a distinct disadvantage, as the heavier outlines would generally tend either to increase or decrease the actual dimensions of an object drawn, which in that case could not be correctly scaled off, or might tend to confuse small sections. With sufficient dimensioning, however, a system of shade lines could very well be employed, which would give many drawings of

that sort of finished appearance, in which respect they are sadly lacking in most cases. It would furthermore assist the eye in reading and interpreting the shape and proportions of parts at a glance. Many offices prohibit outline shading. In such cases all full outlines should be made heavier than others.

Outline shading should universally be employed on all assembled drawings and on such sheets where it will not interfere with scaling off of dimensions or with black sections. Many draftsmen who are in the habit of using shade lines do not make them conspicuous or decided enough. Shade lines should, as a rule, be made about three or four times the strength of the ordinary outline.

The generally accepted rule in regard to shading is to have the light fall from the upper left hand corner at an angle of 45 degrees. The shaded side, therefore, would be the diagonally opposite one, or in a square or oblong figure the side denoting its base, parallel with the lower edge of the drawing and its right hand side. A

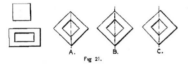

Fig 21.

square opening within any of these figures therefore will have to be shaded the reversed way, or at the top and left hand side. Assuming the square to be set on one angle, as shown in Fig. 21A, only the under side, which is located at right angles to the assumed rays of light, would have to be shaded, while the opening within would have its opposite side so treated. At the same time, as the reader will perceive, something seems to be amiss with that shading. Let us therefore deviate just a trifle from the orthodox rule and apply the remedy: Add one-half or one-third the regular width of a shade line to the lower left hand outline of the square and the upper right hand side of the opening and the drawing ("B,") as well as the one showing the square tilted slightly towards the left ("C") and treated in a reversed manner, appears finished correctly.

The shading of circles is generally effected by "shifting centers" or moving the points of the compasses a trifle from the center of such figure, the distance being equal to the thickness of the desired shade line, and inking in the shading always in a direction from left to right. If the new center is correctly located the shade line merges perfectly at both ends. Another method, as practised by the writer, is to retain the center, shade the arc of a sector of 135° as at "A" and "B" of Fig. 22, the ends of such fully shaded arc tapering and merging into the light portions as Tapers A', A", B' and B" on the same figure indicate. The improvement gained by following the latter method is made fairly clear by comparing "A" and "B" of Fig. 23.

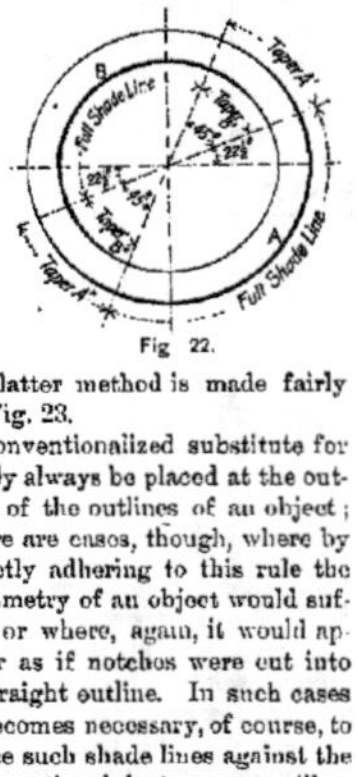

Fig. 22.

Since the shade lines represent a conventionalized substitute for actual shadows cast, they should nearly always be placed at the outside of the outlines of an object; there are cases, though, where by strictly adhering to this rule the symmetry of an object would suffer, or where, again, it would appear as if notches were cut into a straight outline. In such cases it becomes necessary, of course, to place such shade lines against the inner edge of the outline. Such exceptional instances are illustrated and contrasted in Figs. 24 and 25. A striking instance,

Fig. 23.

which illustrates the advantage of shading the outside of objects is shown under Fig. 26.

Very often a case may occur, that rivets, bolts, etc., are to be shown in elevation, where the surrounding parts are cut away or

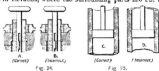

A. (Correct)   B. (Incorrect.)   C. (Correct.)   D. (Incorrect.)

Fig. 24.   Fig. 25.

represented as being in section. Formerly all such objects were uniformly section-lined and consequently not shaded, but that practice has now become obsolete.

Of course the adjacent parts of such a section, as for instance the three thicknesses of plate shown in Fig. 27, are cut on the same

Incorrect.   Correct.

Fig. 26.

plane, and therefore cannot receive any shading between them. The plan "A" on this figure explains why the rivet shown in "B" must be shaded as it is, i. e., the heads overhanging and casting a shadow, and the stem also projecting half way, thus receiving a shade line. The very common mistake committed by draftsmen is to shade them only partially as is shown under "C" in the same figure.

Ordinarily the strength of full shade lines should be made uniform. Imagine for an exception, however, a simple side elevation of a plate girder. If the full-strength shading were given to the overhanging top cover plate of the upper flange it would entirely fill up the thickness of the horizontal flange of the top angle. In such a case a reduction in strength of that shade line to about one-third regular width is advised. The lower edge of the flange of the angle mentioned, outlined against a comparatively wider space, may receive the full strength of shading. Another instance, if such may be quoted, is the shading of the ropes in a drawing of a derrick, perhaps, which shaded full strength would appear too clumsy. Such distinctions are best to be left, however, to the individual taste of the draftsman. In Fig. 28 straight, tapering shade lines are employed; here are indicated in a graphic manner the slanting surfaces of the ribs of a manhole-cover frame. Under certain conditions the actual shadow thrown by this object would correspond with our outline shading shown here. The tapered shading of curved outlines, as shown by Fig. 29, is governed by principles similar to those given for circles and circular openings. The eye, when properly trained, will be able to determine the point of merging from the tapering shade to the light outline.

The shading of a straight line is generally effected by placing the edge of the rule a trifle below it, the pen is opened slightly wider, and

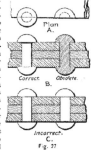

Plan A.

Correct.   Obsolete.

B.

Incorrect.

C.

Fig. 27.

a line ruled parallel to the one to be shaded; the pen is then inclined somewhat towards the draftsman, and in this position will usually fill the space between the two lines by one application. If it is found sometimes that the pen fills the space to be inked by a narrow margin, a slight turning of the inclined pen towards the right between the thumb and the index finger will widen the resulting ink line and make it fully adhere to the two lines, thereby filling the space. Where a horizontal shade line is to be joined by another shade line at any given angle, run the first line a trifle beyond the joint, as shown at right side of Fig. 30. The vertical shade line may then be joined to the extreme end of the horizontal one, and obviously a clean sharp corner in the shading, difficult to obtain otherwise, will

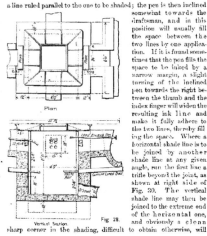

Plan

Vertical Section.

Fig. 28.

result. For shade lines of medium strength the pen is set a trifle wider, and the shade line ruled so close to the outline as to make it adhere there. The corners in such a case need special attention. In shading the straight tapering lines shown in Fig. 28, the edge of the rule is set parallel to the outer edge of the intended shade line and the heavier end is generally commenced with. For circle shading, the full shade line (see Fig. 31) is first filled in with the pen of the compasses in the same order as given for the straight line shading (i. e., draw a parallel line below, in this case a concentric arc, the space between to be filled in by a second or third application.) The tapering ends, as indicated by dotted lines in the figure, are then put in freehand, by beginning in either case from the ends of full lines and working the ink carefully, while yet moist, towards the merging points. This, after a little preliminary practice, can very neatly be done by the average draftsman with the ordinary No. 303 Gillott's pen. A better method, although one requiring more practice, a deft hand and thorough knowledge of the instrument, is illustrated diagrammatically in Fig. 32. Here the shading, as indicated by dotted lines, is effected by gradually springing the compasses or preferably bow

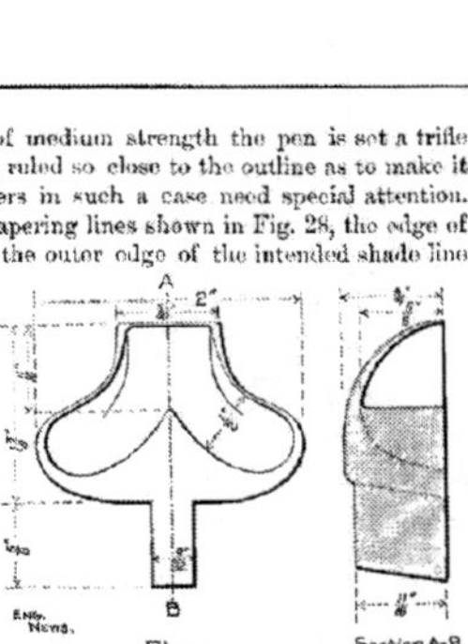

Plan.   Fig. 29   Section A-B.

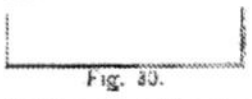

Fig. 30.

pen, from merging point to merging point; a slowly increasing pressure at first, maximum pressure for full portion and gradually decreasing pressure either from or towards the center, as the case may be, will produce in perhaps two applications the strength desired. The relative intensity of the required pressure is indicated on our diagram by differing lengths of arrows. As mentioned, a certain amount of practice is necessary before this method can safely be employed upon an actual drawing. The shading of portions of circles is to be effected also by springing the compasses until the heaviest part becomes equal in width and merges into the adjoining shaded or light lines (such as "A" or "C" in Fig. 33).

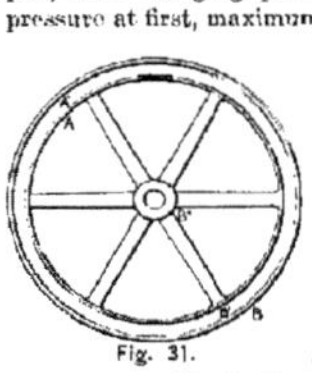

After a drawing has been finished up to this stage (outline shading and lettering completed), the proper section lining and, if desired, the graded line shading, indicating convex or concave surfaces, or, on a map, mountain shading and water lines,

Fig. 31.

Fig. 32.

Fig. 33.

may be added. That style of finishing should, however, never be allowed to run across any lettering which has previously been placed upon such surfaces, but should be broken off, leaving minute open spaces around the letters, as shown in Fig. 34. In this manner the

Fig. 34

descriptive matter is not interfered with in any way. The trifling additional labor which the taking of this precaution entails, is well repaid by the clear and tidy appearance of the respective parts afterwards.

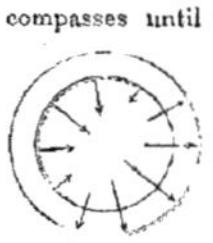

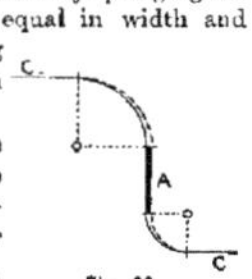

CHAPTER V.   SECTION LINING.

SECTION lines should always be employed where it is desired to represent any object as cut by a plane and where the parts intervening between the observer and the plane of section are removed. An omission of the proper sectioning or "hatching" will result in an unsatisfactory drawing, and will sometimes make it entirely incomprehensible, as the reader of such drawing in trying to interpret it has mentally to supply the missing tints, a very trying and at times irritating task. Some draftsmen again, on the other hand, are in the habit of tinting or hatching all materials represented in the drawing, making no distinction between surfaces in section and such as are in elevation, a proceeding which, entailing as it does, a considerable amount of unnecessary work, will make matters even worse, as far as the legibility of a drawing is concerned. It is of course well to make some concessions in this matter, as for instance, a few courses of brick or stone may be suggested on an elevation of a wall, so as to give relative size of adjoining objects, or to show the kind of bond used, etc. This point will, however, be treated more in detail further on.

The color tinting of sections employed in some establishments instead of the pure black section lining allows, when consistently placed, of a very good mental picture of the part so treated, and the tints will readily suggest themselves: neutral tint for cast iron, blue for wrought iron or steel, yellow for brass, brown for earth, Indian red for brick, etc. When used on a tracing, however, the different colors cease to be distinct on a blue print made therefrom, with the sole exception perhaps that the actinic value of some of the colors may render those more prominent than the rest. For photo-reproductive purposes in black and white a tinted drawing is also useless, as some of the washes would reproduce as a muddy gray or black, others be entirely lost. Therefore it is the writer's opinion that all drawings for photographic reproduction should be finished up in pure black.

In drawing and finishing a section of any object the draftsman very properly should assume the "near" portion cut off by the plane of section as being actually detached; the "far" portion thus remaining should be treated solely and faithfully under this assumption, hatching all such parts as are cut by the plane mentioned, and leaving the unaffected ones severely alone. These few points cannot be emphasized too strongly, as very many draftsmen are sadly deficient in consistently following out such conditions

20

assumed. An almost pathetic instance of erring in that respect is presented in our Fig. 35 "A," and the proper treatment of the parts involved is demonstrated in "B" alongside.

Before introducing the standards of section lining defined below, it may be stated, that in cases where no mistakes regarding different materials employed are possible, or where names of materials are lettered on respective surfaces, there cannot exist any reasonable objection to the use of a simple diagonal hatching across all portions which appear in section. In fact, this method sometimes possesses advantages of its own.

Fig. 35.   A.   B.

The standards of section lining presented in the following have been employed and in parts evolved by the writer while in the drafting department of "Engineering News;" the metal sections given are almost identical with the ones used by the Bureau of Steam Engineering of the U. S. Navy Department. For the materials most often recurring the simplest method of section lining has been chosen; the relative density and probable texture of the materials has also been taken into account. If, while using the standard metal sections given here, the draftsman should happen to be in need of an additional one, as, for instance, phos-

phor-bronze, he may, by adding an appropriate set of lines (perhaps vertical) to the symbol given for bronze, evolve thereby a new distinction and still indicate its relationship, or he might prefer to use the symbol given here and letter on the modified name. In Fig. 36 the chilled contact faces of Car Wheel Iron are indicated by double and triple hatching, so as to show relative density of those materials. Similar instances might be quoted, where a like expedient can be adopted. For all ordinary purposes, however, the writer deems the tables given complete in every respect.

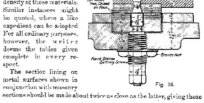

Fig. 36.

The section lining on metal surfaces shown in conjunction with masonry sections should be made about twice as close as the latter, giving those a somewhat darker tint and rendering them distinctly "metallic."

In drawing for ordinary building or bridge construction, the accepted practice is to show all metal sections in black as in Fig. 37. The lines between the different thicknesses of sections are here indicated by faint white spaces, which should, where such

Fig. 37.

thicknesses continue purely in elevation, be likewise continuations of the black dividing lines, somewhat tedious to construct, since a white space between such thicknesses represents two separate outlines. In drawings also where a simple principle of construction, such as shown in Fig. 38, (oil-filter) is to be illustrated, thin metal sections are very appropriately shown in solid black. In all such cases, however, care must be taken never to let the shaded outlines interfere in such a way that they might be mistaken for black sections. For this purpose shade lines of one-half the regular strength may be employed, which may even be thinned down where they occur in proximity to the black sections; or other expedients may be adopted in such a case; that is, the sections may either be slightly increased in width, or shading be left off altogether.

The distinctive metal sections, shown on plate I., and the different masonry sections on plate II., may be executed mechanically—with section liner and ruling pen. From past experience the writer has, however, lost faith in all mechanical devices for doing this class of work—none work satisfactorily, and they cannot supplant the steady hand and alert eye of the trained draftsman. A unique little home-made affair, described by L. F. Rondinella in his paper "Rapid Methods in Instrumental Drawing," is certainly cheap. In the following the description is given in his own words, and the illustration, Fig. 39, attached:

"A very good device, and one that has long been used in some drafting rooms, can be made out of a soft-wood straight-edge, about one-eighth inch thick (e. g., a penny ruler) and two pins, to be used with a triangle and against a T-square blade. One side of the triangle is placed against the upper edge of the soft wood on the paper, so that an adjacent side forms the angle desired for the hatchings, and the pins are driven into

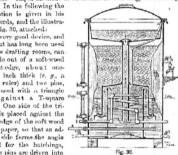

Fig. 38.

the edge so that the corners of the triangle can strike against them, the distance between the pins being equal to the side of the triangle plus the desired distance between the hatchings. To use this device, the lower edge of the soft wood is placed against the upper edge of the T-square blade, as shown in the figure. With the triangle against the left-hand pin, a line is drawn along its right hand edge; the triangle is still held firm, and the straight edge is slid to the left until stopped by the right-hand pin; the straight-edge is then held firm, and the triangle slid up to the left-hand pin; a second line is then drawn, and this process is repeated until the section surface is covered with equidistant lines. After a little practice, work can be done very rapidly with this device, and the eyes are not strained to judge the distance between the lines."

Another simple section-liner, devised by A. S. Burgess and shown under Fig. 40, seems to be an improvement on the above. By changing the triangle in the three slots, we can get a combination giving t w e l v e different widths of spacing, enough almost to cover every case in practice. Section lining by eye requires practice and a steady hand; avoid at the beginning of the work too close a spacing, an error which very many draftsmen commit; start with the ordinary outline strength in a 45° diagonal direction with moderately large spacing; after every 10 or 12 lines filled in glance back over the completed area, so as to gradually correct any deviation towards narrowing or widening spaces. If you notice that a line just ruled is spaced too far from the last one, rectify by purposely ruling the next line too near, and vice versa, assuming immediately afterwards, however, the normal width of spacing. In this manner an even "coloring" of the section-lined surface can be maintained. This is made clear by referring to Fig. 41, showing an evenly hatched surface, where nevertheless "a" and "c" indicate too narrow and too wide spaces respectively, which are rectified by subsequent spaces, "b" and "d," as related above. For very rapid work, the writer employs the medium sized 45° triangle, using its hypothenuse as working edge; the 90° point is raised and the left thumb placed under same, while the four remaining fingers of that hand slightly extended, serve moderately to press the ruling edge down; the triangle, sliding only on the lower edge of its hypothenuse, responds to the slightest motion. The parallel position of the lines, of course, becomes more or less problematical; for the writer believes, while fully realizing the consequences of being considered a confirmed heretic, that truly parallel lines are not essential in section-lining as long as an even tone of the hatched surface is obtained by the means outlined above.

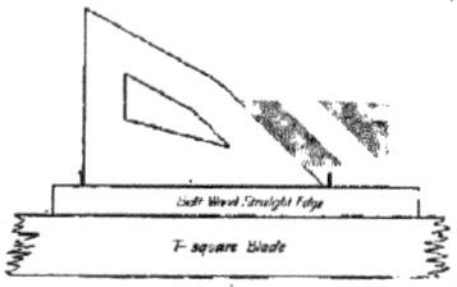

With small or narrow areas of sections the author takes the liberty of doing them freehand altogether, and there is no reason why a draftsman possessed with a steady hand cannot neatly cross-hatch a narrow strip with an ordinary pen. For do-

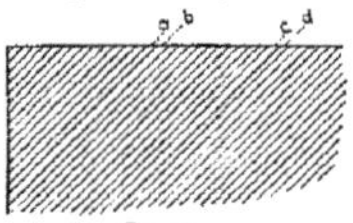

Fig. 39

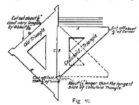

Fig. 40.

Fig. 41.

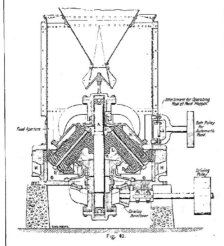

Fig. 42.

ing this class of work on transparent material the detached sheet of "Reinhardt's Lettering" placed under that part of the drawing may be found quite useful in giving spacing and direction of lines, etc. Adjacent areas should receive section lines in opposite direction; where, however, the section-lining of three or more areas adjoins, 30° ruling may in addition be resorted to, as shown by Fig. 42. It is, of course, plain, that the same pieces of metal must here be section-lined in the same direction (as on both sides of vertical shaft "A") so as to establish their identity.

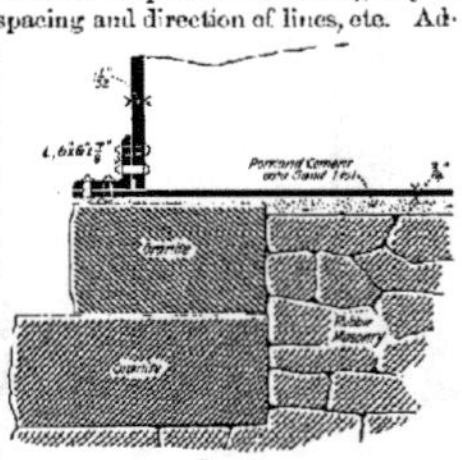

In regard to section-lining brick or stone, as on plate

Fig. 43.

II., the writer would promulgate another pet heresy: these materials never show up so well as when a narrow strip against the lighted sides of the respective areas is left open, thereby enhancing the effects of light and shade; the time expended in ruling in the necessary pencil lines defining such spaces is well paid for by the improved appearance of the sections afterwards. The section-lining for stone, as the denser material should, as shown, be ruled with pen set for about one half the strength of a shade line. Still further distinction in this respect may sometimes be indulged in, as our Fig. 43 shows. Here the

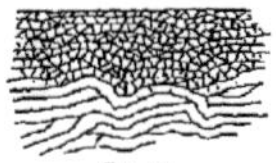

Fig. 44.

light effects on the edges of the separate stones are perhaps uncalled for; they might have been produced at the top and left hand side only, the masonry being thus treated as a whole. In the case of coursed masonry the separate courses are generally not indicated, although in some instances it may be desirable to do so. In "broken stone" and "coursed" (plate IL) the separate stones should be shown angular in shape. Have the rip-rap stones neatly overlapping. Broken stone can be shown by placing each stone separately. When it is desired to show same tightly packed, as for ballast in a roadbed, another method, as shown in Fig. 44, may be employed:

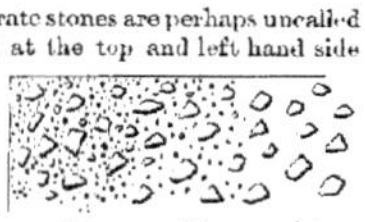

3d.   2d.   1st.

Broken Stone.   Gravel.

Fig. 15.

and then put in cross-strokes at different angles. The stones and gravel in "concrete" should always be shown shaded; the outlines can neatly and quickly be constructed with a single continuous stroke and varying pressure of the pen, as shown in Fig. 43. Throw in the stones at random; then fill in sparingly between these with the heavy, irregularly shaped dots, occasionally putting in a smaller stone where needed. Finally even up the tint of the whole by placing the small dots, indicating particles of cement, which will, when judiciously placed, supply the necessary finish to the section, and make it resemble the real material pretty closely. In "gravel concrete" the gravel should of course receive rounded outlines, and can be quickly constructed, as shown. For the placing of the heavy black dots in "cinder concrete" a heavy lettering pen may be used; each dot to be produced by a single application.

In all these and the following freehand sections the tint should be carried closely against the "light" outlines, unless some other dark tinted section is joined at

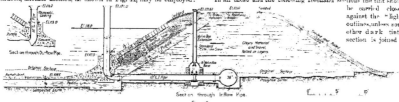

Fig. 45.

Draw sinuous courses, sometimes parallel, sometimes converging, the distance generally being equal to the desired size of the stones, those sides, when narrow open spaces should be left, analogous to rules to be observed in making masonry sections. (Observe "loam" and "puddle" sections against "filling" and "crushed stone" sections, etc., Fig. 45).

The increasing use of reinforced concrete in modern constructive work has necessitated a modification of the standard employed for showing ordinary concrete. In such sections the reinforcing system or skeleton of rods and wires should be shown by fairly heavy, short, double or single dashes, where in side elevation, and by solid black sections, giving circular or square shape of such wires or rods, where this reinforcement appears in cross-section. The reinforcement is thus allowed to stand out bold and clear; the concrete itself is represented in light, broken, diagonal lines, as our Fig. 47 shows. The advantage of using this modified standard for concrete section is demonstrated by comparing A and B, Fig. 48.

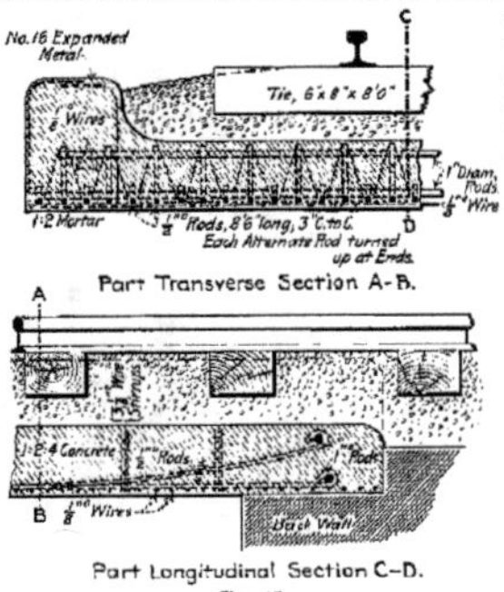

No.16 Expanded Metal.

Tie, 6'x 8'x 8'6"

Wires

Rods, 8'6 Long, 3"C to C. Each Alternate Rod turned up at Ends.

1/2 Mortar

Part Transverse Section A-B.

B-4 Concrete

Rods

Wires

Side Wall

Part Longitudinal Section C-D.

Fig. 47.

"Gravel" is, wherever occurring in the following sections, never shaded, so as to avoid any of its compounds, such as "sand and gravel," being mistaken for concrete. The symbol for "sand" can

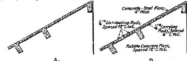

Concrete—Steel Floor, 5" thick.

Distributing Rods, Spaced 12" C. to C.

Carrying Rods, Spaced 6" C. to C.

Rubble Concrete Piers, Spaced 12" C. to C.

A.   Fig. 48.   B.

with a little practice be evenly placed by connecting groups of dotted circular arcs, as shown in Fig. 49.

Water in repose (Plate III.) may be represented in two ways, as the exigencies of the case may demand : by graded solid lines, decreasing in strength as their spacing increases, or by evenly spaced, uniformly broken lines. The latter method is employed for showing "oil" in Fig. 38, and it answers very well for contrasting the two liquids shown in that illustration.

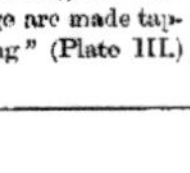

Fig. 49.

Earth is indicated by series of short parallel strokes, about 4 or 6 in a bunch, constantly changing in direction. Where only a narrow strip of an earth section, such as under a foundation, is to be indicated, a neat way of toning it down in color is suggested in Fig. 50, which shows the three stages of drawing the section. As will be noticed, the strokes in the first stage are made tapering downwards and placed closest. "Earth filling" (Plate III.)

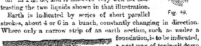

3d.   2d.   1st.
Fig. 50.

can be shown by disconnected parallel-lined bunches interspersed with dots, showing a generally broken-up texture. In drawing the symbol for "bed rock" it will first be necessary to throw in the heavier dividing lines, indicating seams, before the actual filling-in can be commenced. Make those tapering, as shown in Fig. 51, and irregular in direction and length; some extending and breaking up at a little distance out. Where quite a deep section of

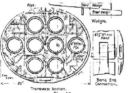

rock is required lateral seams may be put in additionally, as shown. In filling between seams by sets of parallel lines, their direction should change as abruptly as possible between the different sets.

In showing stratified rock, horizontal seams, or nearly so, should be predominant, except in a case where a dip of the strata occurs, when naturally the seams must be drawn so as to represent that inclination. If any filling-in at all is desired here, it may be done sparingly against the edge.

Fig. 52.

The "undefined" symbol might indicate either bed rock, earth or sand; it is with advantage employed where the actual material is not definitely known, and is executed, as shown on plate III., by wavy, freehand lines.

In showing "puddle" section on same plate the horizontal strokes may be done either freehand or by rule; they should however, break joint neatly; the same may be said of the symbol for "silt," where the shorter dashes should be pretty closely placed. "Cement," as distinguished from "sand," is shown with large and small dots. Where drawing some of the geological symbols contained on Plate IV., throw in for instance "boulders" at first at random, "cobbles" or "gravel" afterwards as fillers.

Cross-sections of wood shown on same plate, to be done freehand of course, should always represent circular arcs; where complete "rings" are drawn, have them as nearly circular as possible; increase spacing from the core outwards in slightly wavy lines. The drying cracks or "checks," which are one of the principal characteristics of the transverse wood section, should be made nearly radial; a little critical study of the sawed-off end of a square timber will be helpful. Fig. 52 appended herewith, shows a cross-section of a timber electric conduit, and is a pretty fair specimen in showing the grain of the different portions.

Longitudinal sections of wood are generally drawn by wavy lines fairly evenly spaced. If more refinement is desired, the study of the texture of a planed pine board may be advisable. As shown by Fig. 53, a judicious

Fig. 53.

placing of a few solid blacks by back and forward strokes of the pen, parallel with the axis of the timber, will produce the desired effect. The writer deems it unnecessary to adopt different kinds of graining for sections of various specimens of timber; distinctions in that direction may be attained by lettering on the relative names.

Additional specimens of geological symbols are given on plate V. in "sections of well borings." Note how increase in density, signifying increasing hardness of the "red sandstone" in the first boring, may be represented in such a case.

The portion of "covered filter bed section" on plate V. shows successive layers of filtering materials; different sizes of filter gravel may be designated as shown; observe how the light strip at the base of the two upper layers of that material enhances clearness. The symbol for "sand" is only filled in downwards some distance from the edge for reasons of saving time; the effect produced thereby is better than a carefully worked over whole surface could give. On the "manhole-section" introduced on the same plate every material shows so distinctively that the descriptive matter was totally dispensed with on that drawing. The sections of the vitrified pipe were here, for exceptional reasons, left entirely white.

As the drawings of more recent electrical constructions are as a rule inconsistently treated and therefore hard to understand, the writer

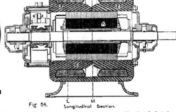

Transverse Section L-L.

Fig 54.

Longitudinal Section.

thought it expedient to insert under Fig. 54 two sections of an electric motor. In ordinary drafting, as well known, no distinction would be made between parts of laminations "A" in section or in elevation. In our drawing this point is strictly maintained. The exact thickness or the number of those plates is of course not given, and, if necessary, such information may be lettered alongside. Insulations are shown black (where in section), although the ordinary electrical draftsman generally puts in all insulating material in that color, whether shown in section or elevation.

The drawing from which our illustration was copied, indicated the actual courses of the wires, presumably, in the circumferential coils "F" in "Longitudinal Section" by horizontal and vertical lines, and in "Transverse Section L-L" the coil "F" showed concentric rings corresponding with the horizontal joints. This arrangement, though perhaps unnecessary, was faithfully copied, so as to make a concession to existing notions upon the subject. In order, however, to show those parts clearly as section, a diagonal set of lines was added, which, although at variance with existing practice, was the only consistent way to treat the portions involved.

CHAPTER VI. CURVED SURFACE SHADING.

THE portions of a drawing which represent curved surfaces may be executed either by graduated tinting (wash) with water colors or by line shading. The latter process, which will be described in detail below, requires, if properly carried out, considerable practice and skill. The style of cylinder shading generally taught and used by some draftsmen, is shown by Fig. 55A, and looks exceedingly neat, where no outline shading is supposed to have been employed. Where such, however, has been used, and curved surface shading is to be added for the sake of clearness, these surfaces must logically be represented as at "B," showing the darkest shading near the edge.

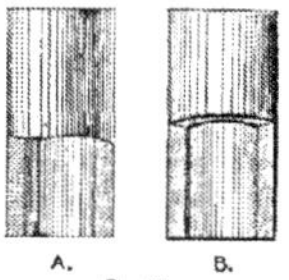

The theoretical method of curve shading is made so clear by Fig. 56, that no further explanation will be necessary. An unusually poor example of such work from a plate of the Mississippi River Commission's report is shown under our Fig. 57, demonstrating to a certainty that the draftsman did not follow such rule, and furthermore, did not possess the redeeming trait of being able to space his lines by eye.

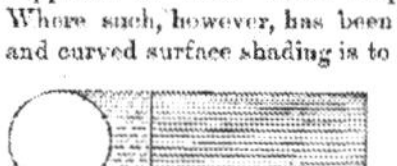

Definite rules for the construction of graded curve shading, fair examples of which are given under Fig. 58, and on plate V. (Manhole-Section), would be difficult to formulate and could not always be applied successfully; therefore the writer will content himself with giving a few concise hints relative to the subject matter for the guidance of the draftsman.

Begin ruling the light side of a cylinder from the edge towards the center with lines of uniform strength; place the first line as close as possible to the outline proper; increase spacing of the lines

Fig. 55.

A.   B.

Fig 56.

Fig. 57.

gradually at about the rate of 3 to 5; keep on increasing spaces until near center, then stop. When ink on lines just ruled has

dried, begin at the opposite or dark side; rule a line about three-fifths the strength of the shaded outline, as near as possible, leaving just a trace of a white space between the two. The next line ruled should

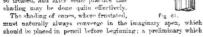

Fig. 58.

again be about three-fifths the strength of the preceding one, with spacing as on the light side, a trifle wider between. Keep on in this manner until no more grading of lines is possible, the minimum or outline strength of line having been reached, when a similar increase of widths of space may be resumed. End near or

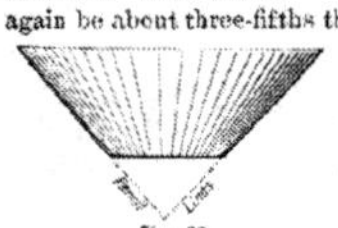

Fig. 59.

at the center line, leaving an open space against the other side. Carefully watch the respective grading and spacing of the lines, also make sure that the open space which represents a high light on the curved surface, comes to one side of the center line of the object so treated, and after some practice this shading may be done quite effectively.

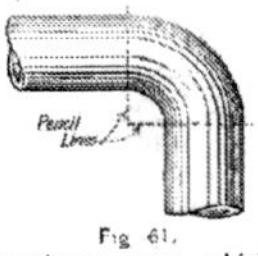

The shading of cones, where frustated, must naturally always converge in the imaginary apex, which should be placed in pencil before beginning; a preliminary which

Pencil Lines

Fig. 61.

many draftsmen deem superfluous, and the omission of which will always be noticeable. In shading the dark side with graded lines, care should be taken, for obvious reasons, to have those shade lines tapering toward the apex, as our Fig. 59 shows. It will help the clearness and also improve the general appearance of the drawing, if a white edge like that described for "masonry sections" in the preceding chapter, is left open against the light side of the ends of the curve

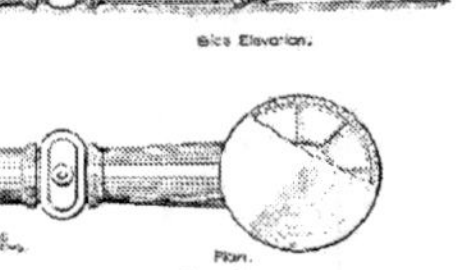

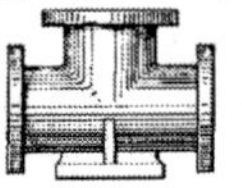

Side Elevation.

Fig. 62.

Plan.

Fig. 60.

shading, a treatment which all illustrations accompanying the present chapter have received.

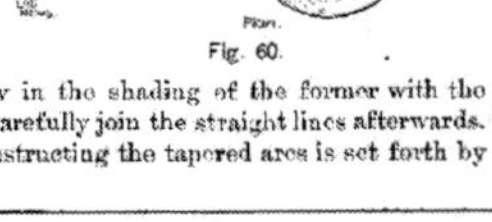

Where a curved portion adjoins a straight part, as for instance shown in Fig. 60, the best way to proceed is to throw in the shading of the former with the compasses first, and then carefully join the straight lines afterwards. The proper manner of constructing the tapered arcs is set forth by our Figs. 23 and 32, which accompany the chapter on outline shading. Fig. 61 illustrates the use of pencil guide lines to insure a perfect joint at the points of tangent. It sometimes may be desirable to increase the shade effect of pipe flanges, for instance, in which case freehand shade lines, as in Figs. 60 and 62, parallel to the flange in this case, may be judiciously placed.

As the draftsman may sometimes be pressed for time, and some curve shading should nevertheless be employed for the sake of clearness, the handy expedient of only shading curved objects on the dark side may be resorted to, as shown in our illustrations, Fig. 63,

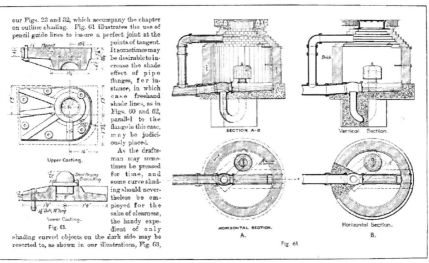

Flanged
Upper Casting.
Steel Forging
Bronze Ring
Lower Casting.

Fig. 63.

Brick
SECTION A-B
Vertical Section.
HORIZONTAL SECTION.
A.
Horizontal Section.
B.

Fig. 64

and the drawing of the D'Auria Pump on plate VI. This leaves the eye to supply the missing shading, and answers well. The little flat freehand shading on the latter figure used sparingly helps in this particular case to give flat effects against the curved portions, where left blank.

Piles may be shaded very neatly freehand with a 303 Gillott's pen in the manner shown on "Standard Pile and Trestle Bridge," plate VI. The strokes composing the shade lines should be made wavy, so as to represent a somewhat rough surface. Where de-sired, this shading may also be executed on one side only. While going to some extent into details for shading on curved surfaces, the writer would employ this kind of finish only where clearness demands it, or where it is essential that a highly finished draw-ing be produced. He would certainly deprecate the use of curve shading in connection with a drawing where the essentials of correct drafting are so flagrantly violated, as on Fig. 64 "A," and would only use it where absolutely necessary, as shown in the cor-rected drawing, Fig. 64 "B."

CHAPTER VII. SHADING OF INCLINED SURFACES.

INCLINED surfaces are, as a rule, represented by parallel, evenly spaced lines; the greater the angle of inclination of any surface from the picture plane the closer the spacing of those lines should be. This kind of shading should again only appear on mechanical drawings for special purposes, or where clearness demands it.

Inclined light surfaces should receive light line shading, while the dark inclined spaces should be ruled by lines of similar strength as the shaded outlines defining such area. This is suffi-ciently made clear by the illustration on plate VII., where in "Plan of Invert" the light stepped-off incline shading joins the dark curve shading and vice-versa. This shading has to be very carefully done where the incline becomes tangential to a curve, as shown for instance on "Hinge Support Casting" on same plate, and the relative spacing can here be determined by dividing the total inclined or curved surfaces into equal parts, and then projecting those points from the elevation to the plan. The suc-cessful draftsman will of course sooner or later depart from the strict rule and depend solely upon his eye to determine the proper spacing.

Where desired, this kind of spacing can be sparingly applied, as in the case of the curve shading; that is to say, light surfaces may be left blank altogether, as the plan of the "Funnel-Shaped Spillway" on plate VI. indicates. The plan of the "Reservoir Embankment" on plate VII. shows purely freehand lines parallel to top and foot of slope, to denote a rough surface. Such slopes may in some instances also be indicated by tapering, wavy strokes, drawn at right angles or radial to the top, and

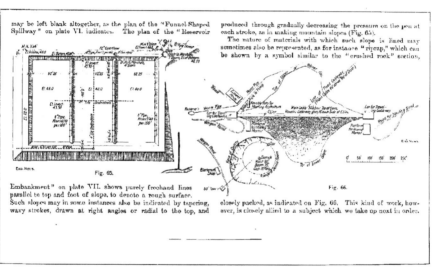

Fig. 65.

Fig. 66.

produced through gradually decreasing the pressure on the pen at each stroke, as in making mountain slopes (Fig. 65).

The nature of materials with which such slope is lined may sometimes also be represented, as for instance "riprap," which can be shown by a symbol similar to the "crushed rock" section, closely packed, as indicated on Fig. 66. This kind of work, however, is closely allied to a subject which we take up next in order.

CHAPTER VIII. TOPOGRAPHICAL DRAWING.

THE mechanical draftsman may at times be called upon to execute maps more or less sketchy in appearance, this kind of work cannot altogether be styled mechanical drafting. In the following the writer will endeavor to limit himself almost exclusively to the subject of "sketch maps;" once proficiency in such work is acquired however, by the average draftsman, he will be able to produce nearly or quite as good work as a regular topographical draftsman can accomplish.

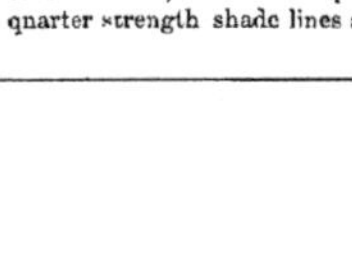

Freehand outlines should always be drawn with a medium strength pen; a well-worn Gillott's No. 303 will answer nicely. Outline shading of any map is essential; shade rivers, roadways, streets, lakes, etc., strictly as depressions, while city blocks, bridges, buildings, outlines of land against water and similar objects should be treated as projections. The freehand shading of coast lines, or of roadways must be done punctiliously; half or quarter strength shade lines should be used at the proper angle, as

Fig 67

has been set forth by Figs 29 and 33, and is indicated here by Fig. 67.

The features which are shown upon any map may be classed in

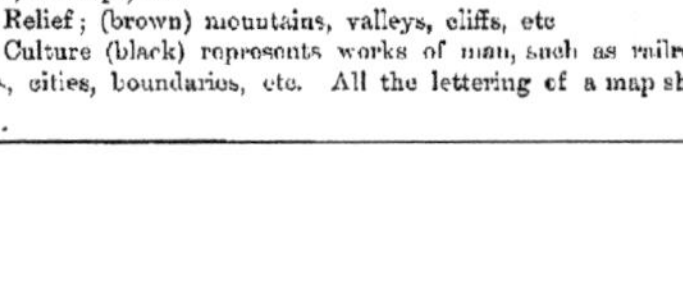

Fig 68.

three groups, and where coloring is to be employed may be indicated by suitable tints, as

1. Water; (blue) represented by seas, lakes, ponds, streams, canals, swamps, etc

2. Relief; (brown) mountains, valleys, cliffs, etc

3. Culture (black) represents works of man, such as railroads, roads, cities, boundaries, etc. All the lettering of a map should

also be done in black. The built-up portions of cities can be shown in solid black, as also the single houses represented.

Another group may here be mentioned. The vegetation, as woods, meadows, orchards, etc., may be indicated by easily understood conventions and tinted light green. Where coloring is not desired these different features may, of course, be shown in black, and are also represented here in that manner under Figs. 68, 69, 70 and 71.

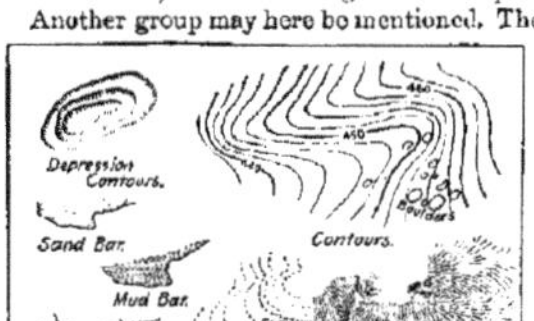

Fig. 69.

After the necessary outlines of a map have been inked in and the outline shading has judiciously been placed, the lettering is put on. The style of lettering to be used on sketch maps is described in the

author's work on "Lettering," and is deemed more in keeping with the general style of such maps than the orthodox roman and italic letters. The subsequent work consists in finishing up proper.

Watercourses, where indicated by a single line, should be drawn

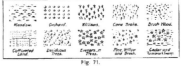

Fig. 71.

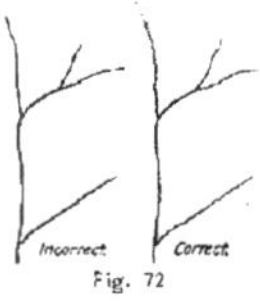

in a direction towards the source, so as to end there with as finely tapering a stroke as possible. Avoid the flat, meaningless style of line, as under "Incorrect" in Fig. 72, but endeavor to produce a wavy or "wabbling" effect, as shown alongside. Where, however, two opposite shore lines are used to denote the width of a watercourse, or where in general, water adjoining land portions is to be shown, thin lines running along the shore, similar to contour lines, may be used to indicate such water surfaces. Draw the first one as close as can possibly be done all around the land portions. Follow with the subsequent ones, finishing each contour completely before touching the next one. Observe increase in spacing, which should be about the same as advised for light side in chapter VI. "Curved Surface Shading." Take care to have such lines pass under a bridge for instance, and around an island or pier, as exemplified by Fig. 73. Closely follow at first the shore line into every nook and corner; the next contours will, of course, gradually develop into more or less flat curves. The writer uses for the drawing of such "water-lines" a Gillott's Lithographic or mapping pen. Good examples of such water shading are given under "Map of Porto Rico" and "Location of White Pass and Yukon Ry.," on plate VIII. About 6 to 10 lines will be found amply sufficient to offset the land neatly. Such work, when properly executed, gives an excellent finish, pleasing to the eye and the lines suggest the form and relative spacing of actual waves coming shoreward.

Fig. 73.

When it is desired to spend somewhat less time and still attain a suitable finish for the land portions, they may be offset against the water by adopting the shore-hatching. By this method uniformly short, evenly spaced lines, drawn in a direction as nearly as possible radial to the curves of the shore line, are used. Fig. 74 explains the principle involved very well. This treatment gives a most realistic effect on an indented, broken-up coast line. In the map of the "North Sea Countries of Europe," plate VIII., this hatching has exclusively been used.

Fig. 74.

A sandy flat coast can be offset by the method shown in Figs. 75 and 76. In the former illustration the rows of dots are placed in a similar manner as the "water-lines," the spaces widening out inland. The sand dunes on Fig. 76 are also similarly treated.

Contour lines on a map, (Figs. 69 and 77) must be indented by very light lines. Every fifth or tenth contour should be accentuated by drawing same heavier. For this purpose a well worn Gillott's No. 303 pen may be recommended; this will yield a fairly uniform line of medium strength, requiring no pressure when once the ink has well started flowing.

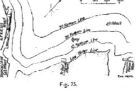

Fig. 75.

The convention for "Bridge" in Fig. 70 may be modified to suit conditions, as for instance a small plan of a drawbridge with center pier, etc., may be substituted for the type shown in our illustration. The strokes indicating slope in "Cut" and "Fill," or "Levee" in the same figure, should be drawn in a "wabbling" or vibrant manner, and, beginning with heavy pressure, gradually decrease in strength until they end in a hair-line (see Fig. 78). They should always be drawn in a direction towards the body. A good draftsman with a trained eye

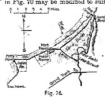

Fig. 76.

and steady hand, however, may construct these strokes in a reversed way, beginning with hair-line strength and finish with the heavy stroke.

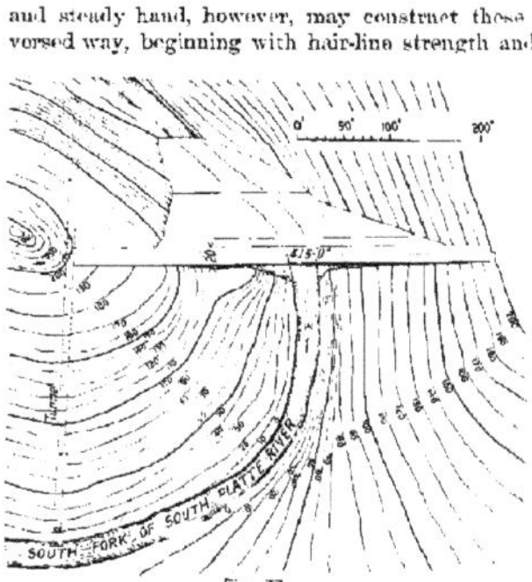
0'
50'
100'
200'
SOUTH FORK OF SOUTH PLATTE RIVER

Fig. 77.

In Fig. 79 is shown an instance where, for specific purposes, a good many distinctions for railways had to be evolved; this table or legend is taken from a railway map published in "Engineering News."

The conventions shown in Fig. 71 can be supplemented considerably by adopting suggestive, easily understood symbols for different kinds of crops. The tufts of grass in "meadow" should begin and end in short, light strokes of the pen; the correct and incorrect ways are shown in Fig. 80A. The distinction between "orchard" and "wood" is so obvious that no further description is necessary.

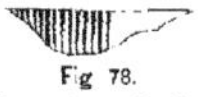

Fig 78.

The method of easily and correctly forming deciduous trees is also indicated in Fig. 80, under "B." Make as near circular as possible, a very important point. "Evergreen" trees are shown by five-pointed asterisks, of which the first three strokes, as in "C" of the same figure, are each placed by one application of a pliable pen at or nearly the correct angle, so as to leave room for strokes 4 and 5. The conventions for "willows" are to be done in a similar manner as the arrows for dimension lines. In "Brush Wood" small trees are drawn and tufts of grass placed at intervals. The symbol for "Cultivated Land" may also be used for fallow or freshly plowed land, and on a color map yellow ochre be used instead of green.

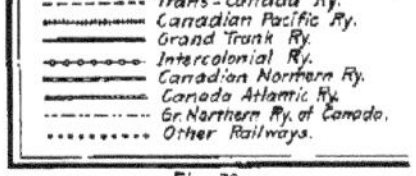
Grand Trunk Pacific Ry. or National Transcontinental Ry.
Trans-Canada Ry.
Canadian Pacific Ry.
Grand Trunk Ry.
Intercolonial Ry.
Canadian Northern Ry.
Canada Atlantic Ry.
Gr. Northern Ry. of Canada.
Other Railways.

Fig. 79.

The mountain work on a map can be executed by a fair draftsman in quite a satisfactory manner after some practice and a little attention to the following directions. The contour lines should first be approximately penciled in from the available notes; the hatching or short disconnected lines, by means of which the mountain shading is effected, should always be drawn vertically to the contour and in a direction downward from the summit (see Fig. 81). Have the strokes overlap slightly at times rather than leave open spaces between the rows of hatched contours (see A and B, Fig. 81). Where contours are not parallel the hatching lines must be drawn radiating, so as to have their extremities normal to the contour at which they terminate.

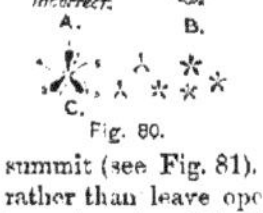
Correct.
Incorrect.
A.
B.
C.

Fig. 80.

The degree of slope is indicated by varying the distance between, as also the thickness of the hatching lines. Such lines accordingly are drawn heavier and closer spaced as the slope is steeper, and finer with wider spaces between for gentle slopes. Where a slope suddenly becomes abrupt the tint must be deepened, or short lines at right angles to hatching must be interpolated (see "Cliffs" in Fig. 69).

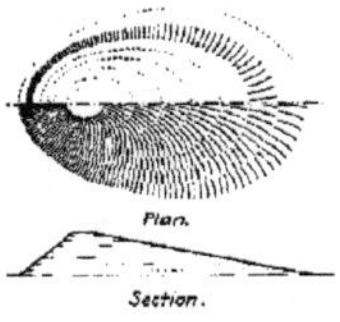

The hatching must therefore run parallel to the "line of greatest descent," or a line indicating the shortest course which water would take running down from the summit. This will account for the curving outward of those strokes in the projecting slope in Fig. 82 "A." A draftsman who has acquired the practice of drawing the long, tapered, single strokes indicated in Fig. 78 satisfactorily, can fairly well shade the slopes shown by "A," Fig. 82 with the single strokes, as outlined in "B," of the same figure.

Fairly good examples of mountain work are contained on plate VIII. and on Fig. 34.

Take a not too fine, well-used pen to execute the hatching with. On the map of the "White Pass and Yukon Ry." the mountain work is treated somewhat sketchily through only finishing one side of a hill or mountain range for instance, as a means of saving time and to prevent an overcrowding of the map, the whole of the country represented being quite mountainous.

Fig. 82.

CHAPTER IX.   CHARACTER AND FINISH.

THIS chapter is principally devoted to supplement the subjects treated upon in the foregoing. Thus there may be cases, where it is desirable to add sparingly a few touches—features which in the preceding chapters have not altogether been enumerated.

In the instance of the "Masonry Portal of Sewer," plate IX., the rock face of the stones can very appropriately be denoted by a few sketchy freehand strokes against the light side of the wall, or some courses of brick may be indicated in the side wall of the boiler setting of the "Playford Stoker" on the same plate, and still not interfere with the clearness of the drawing, the section-lined parts standing out clear against the back-ground.

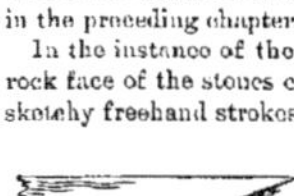

Fig 83.

The few sketchy touches upon the piers of the "Harlem River Drawbridge" (plate XI) have been put in for more than ornamental purposes, and a close scrutiny of the drawing will reveal in a way the nature of the masonry employed, and the different methods of foundation adopted for the piers.

Where wood in elevation is to be distinctly treated, a few wavy strokes, as in "Sectional Elevation of Caisson," plate IX, and in "Standard Pile and Trestle Bridge," plate VI, will in the former instance, where shown in "Half End Elevation," answer the purpose of contrasting the outer sheeting as "wood" against the metal plate of the cutting edge. When it is overdone to such an extent, however, as in the "Caisson for New East River Bridge," plate IX, it will tend to obscure any distinction between the parts in section and eleva-

Fig 85.

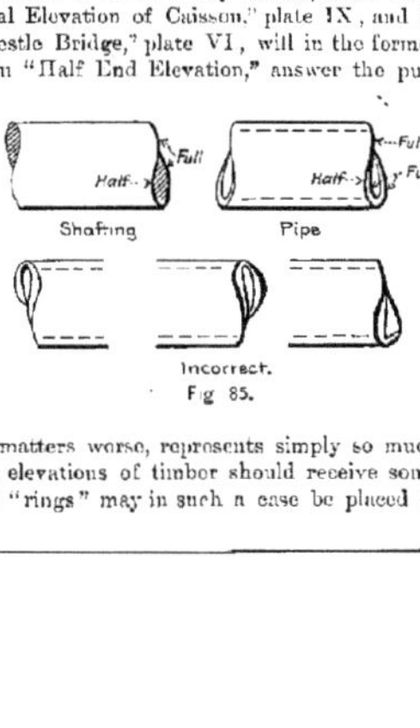

tion, and only making matters worse, represents simply so much time wasted. Pure end elevations of timber should receive some treatment, and parts of "rings" may in such a case be placed in a sketchy way, although the whole area to be shown as end elevation should never be filled in completely, as the upper part of the illustration just mentioned on plate VI, will show.

Such treatment, if judiciously applied, suggests at a glance a number of instructive details, and furthermore enhances the appearance of a drawing. In order still more to heighten such an effect the breaks, if breaking off of materials has to be resorted to, should also be made to suggest the nature of such materials, and even their shape. Thus wood should, as a rule be broken off, as Fig. 83 shows, so as to contrast with metal or stone. In Fig. 84 a few characteristic specimens, showing a formation of breaks, suggestive of the relative shape of the objects, are represented. The consistent following out of the rule given above, seemingly results in the case of the "sewer" on the same figure, almost in a perspective representation of that object, which, perhaps, would be carrying the point

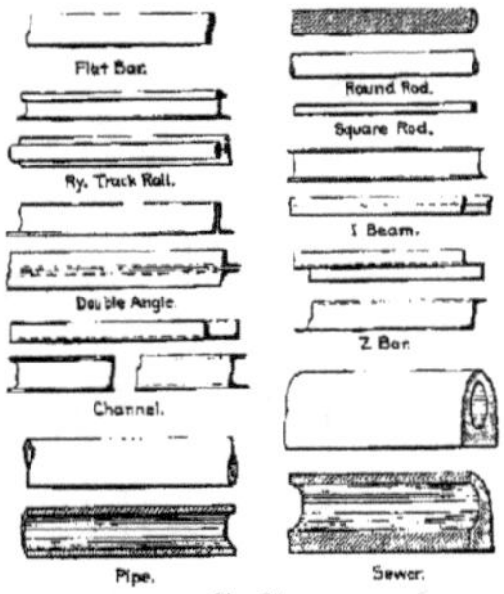

Flat Bar. Round Rod. Square Rod. Ry. Truck Rail. I Beam. Double Angle. Z Bar. Channel. Pipe. Sewer.

Fig. 84.

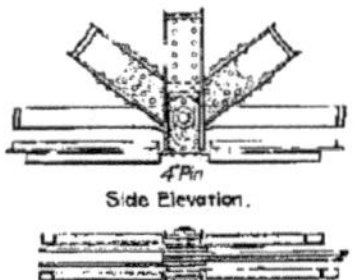

too far. The collection given here can, where occasion offers, be enlarged upon by the individual draftsman.

Just a word or two in regard to the correct construction of such "breaks." First of all, it must not be forgotten, that a very decided distinction should be made between breaks of shafting and of pipe. Under Fig. 85 the correct style of the outline shading of such breaks is demonstrated, showing how half and full strength shade lines can be used, so as to properly indicate the

Side Elevation.

Sectional Plan.
Fig. 86.

shapes. The lower portion of breaks are occasionally drawn and shaded up by the inexperienced; here the draftsman was not able to realize, that the greatest thickness of the pipe wall could consistently only be shown either at the top or at the bottom of the break, and that the same inevitably must decrease in width towards the middle at the sides.

The elevation and sectional plan of the panel point of a bridge span, in Fig. 86, shows

Rivets.

Screw Head. Bolt Nuts. Shaded Outlines only. Rope or Cable.

Fig. 87.

without a word of descriptive matter to a trained eye the shape and "make-up" of the different members composing the structure. The duplicate breaks in the elevation might, of course, be dispensed with, where the number of bars or channels are given in the lettering descriptive of the "make-up."

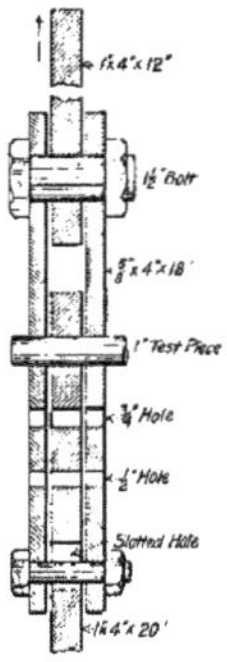

In the representation of a timber crib, on plate XI, the isometric view shows how ordinary outline shading may be applied to a perspective drawing—somewhat at variance with the outline shading of pure plane projections. Here the ends of the timbers are for obvious reasons broken off in a very plain manner.

Rivets for metal structures are, as a rule, left unshaded, except where drawn to a large scale. In the latter case, shading may be applied after one of the fashions shown on Fig. 87. The writer thought it unnecessary to employ any code for the various manners of riveting, as the different bridge and construction concerns have each formulated a standard of their own for such work. Shading of large scale bolt-nuts may with advantage be used, as shown on some illustration and in Fig. 88. On assembled drawings, the writer sometimes goes to the extreme of showing rivets or nuts in plan, as represented again in Fig. 87, outlining only the shaded portions. This method may be recommended where effect alone is desired.

In regard to Fig. 89, which illustrates the rational shading of

Fig. 88.

screw-threads, little, if anything, will be said, as the drawing is self-explanatory. The "full" shading of the triangular threaded screw may be recommended as especially effective, although the ordinary shaded outline is sufficiently distinct for ordinary purposes. Under the sub-caption "threads flat" in the square threaded screw the effect of unshaded threads against the shaded stem is shown. Here the absence of shading in the threads really conceals their correct shape in representing them as straight surfaces, and a little free-hand shading, as shown above under "shaded full," is necessary to bring out the fact that they are circumferential in plan to a round stem. For showing

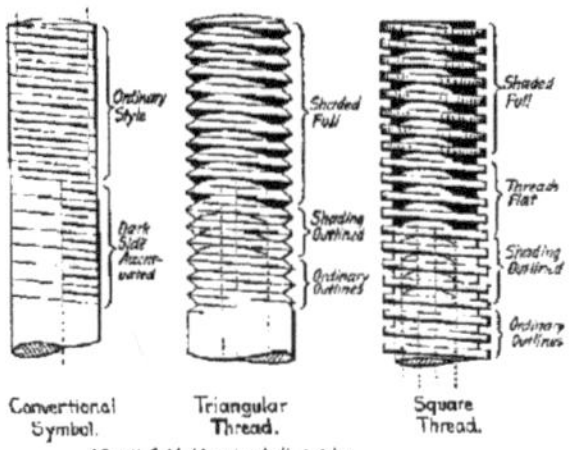

Conventional Symbol.   Triangular Thread.   Square Thread.

(Pencil Guide Lines are indicated by Vertical Dotted Lines)

Fig. 89.

A.   B.

Fig. 90.

41

square threads at a smaller scale, the ordinary outline shading, of course, will suffice. Another method of shading square screw-threads, which is not shown here, is to employ convex line shading, described in a preceding chapter—quite a tedious piece of work, at the whole. The proportioning of spaces of the pencil guide lines indicated on our illustration practically suffices for laying out the curved outline of shading described here.

The correct manner of shading male and female screw-threads in proximity is shown in Fig. 90. The shading of the lower part of "A" is just the reverse of that of the upper portion. The slope of the female thread is also shown in the opposite direction, as the portion which is seen beyond the plane of section corresponds to the hidden part of the (male) thread of the bolt. This matter is simplified or conventionalized in "B" of the same figure; make sure to keep the outline shading of the bolt and the female threading inside, and do not encroach upon the section-lined area on either side.

The elevation of the Columbus, O., Freight Station, on plate XI. demonstrates finally how well a sketchy method of treating masonry, etc., adapts itself to architectural drawings. The treatment of the windows, especially, is well worth studying.

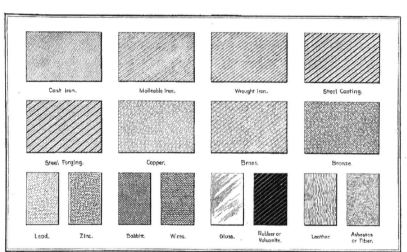

PLATE I.

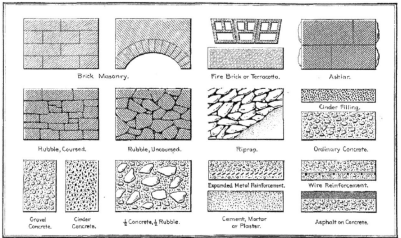

Brick Masonry.

Fire Brick or Terracotta.

Ashlar.

Rubble, Coursed.

Rubble, Uncoursed.

Riprap.

Cinder Filling.

Ordinary Concrete.

Gravel Concrete.

Cinder Concrete.

½ Concrete, ½ Rubble.

Cement, Mortar or Plaster.

Expanded Metal Reinforcement.

Wire Reinforcement.

Asphalt on Concrete.

PLATE II.

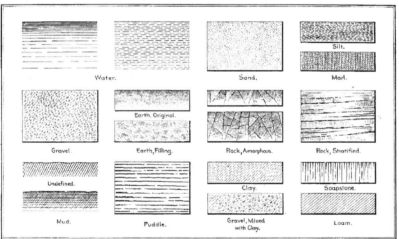

Water.

Sand.

Silt.

Marl.

Gravel.

Earth, Original.

Earth, Filling.

Rock, Amorphous.

Rock, Stratified.

Undefined.

Clay.

Soapstone.

Mud.

Puddle.

Gravel, Mixed
with Clay.

Loam.

PLATE III.

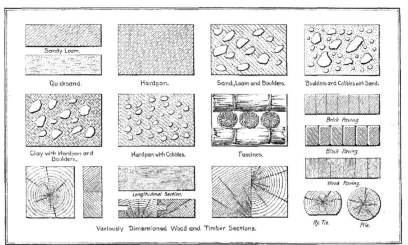

Sandy Loam.

Quicksand.

Hardpan.

Sand, Loam and Boulders.

Boulders and Cobbles with Sand.

Clay with Hardpan and Boulders.

Hardpan with Cobbles.

Fascines.

Brick Paving.

Block Paving.

Wood Paving.

Longitudinal Section.

Variously Dimensioned Wood and Timber Sections.

Ry. Tie.

Pile.

PLATE IV.

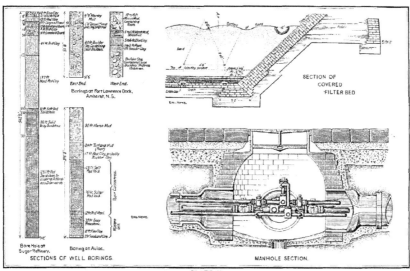

SECTION OF
COVERED
FILTER BED

Borings at Fort Lawrence Dock,
Amherst, N.S.

Bore Hole at
Sugar Refinery.

Boring at Aulac.

SECTIONS OF WELL BORINGS.

MANHOLE SECTION.

PLATE V.

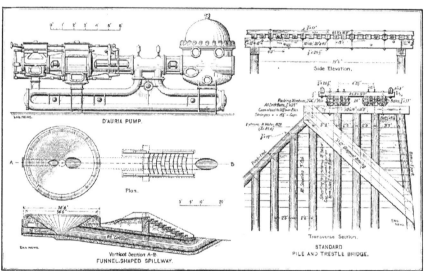

ENG. NEWS.

D'AURIA PUMP.

Plan.

A ———— B

Vertical Section A-B
FUNNEL-SHAPED SPILLWAY.

ENG. NEWS.

Side Elevation.

Transverse Section.

STANDARD
PILE AND TRESTLE BRIDGE.

PLATE VI.

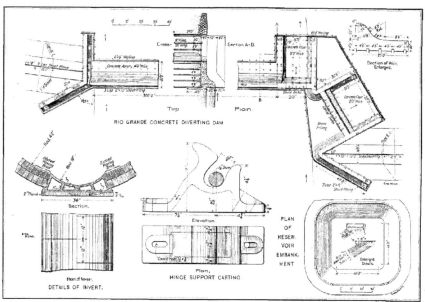

RIO GRANDE CONCRETE DIVERTING DAM

Cross-   Section A-B.

Section of Weir,
Enlarged.

Top   Plan.

PLAN
OF
RESER-
VOIR
EMBANK-
MENT

Section.

Elevation.

Plan of Invert.   Plan.

DETAILS OF INVERT.   HINGE SUPPORT CASTING

PLATE VII.

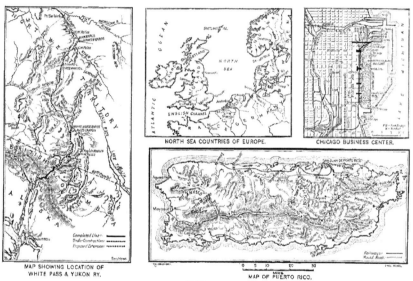

MAP SHOWING LOCATION OF
WHITE PASS & YUKON RY.

NORTH SEA COUNTRIES OF EUROPE.

CHICAGO BUSINESS CENTER.

MAP OF PUERTO RICO.

PLATE VIII.

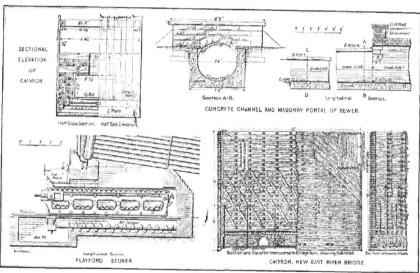

SECTIONAL
ELEVATION
OF
CAISSON

Half Cross Section.   Half End Elevation.

Section A-B.
CONCRETE CHANNEL AND MASONRY PORTAL OF SEWER.

C

D   Longitudinal.   B Section.

Longitudinal Section.
PLAYFORD STOKER.

Section and Elevation transverse to Bridge Axis, showing Bulkhead.   Section showing inside.
CAISSON, NEW EAST RIVER BRIDGE.

PLATE IX

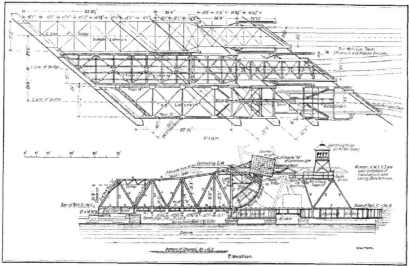

PLATE X.

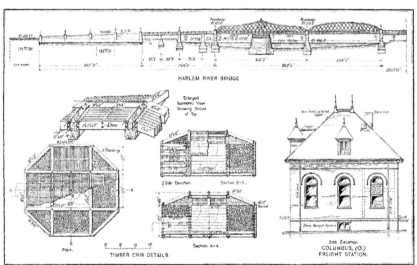

HARLEM RIVER BRIDGE

Enlarged
Isometric View
Showing Portion
of Top

Plan.

TIMBER CRIB DETAILS.

Side Elevation.   Section B-B.

Section A-A.

End Elevation.
COLUMBUS, (O.)
FREIGHT STATION.

PLATE XI.